W9-BNU-726

AMERICAN LANDMARKS

The Barn

AMERICAN LANDMARKS

The Barn

LAURA BROOKS

This edition published in 1997 by SMITHMARK Publishers, a division of U.S. Media Holdings, Inc.
16 East 32nd Street, New York, NY 10016.

SMITHMARK books are available for bulk purchase for sales promotion and premium use.
For details write or call the manager of special sales, SMITHMARK Publishers
16 East 32nd Street, New York, NY 10016; (212) 532-6600.

This book was designed and produced by Todtri Productions Limited
P.O. Box 572, New York, NY 10116-0572 FAX: (212) 279-1241

Printed and bound in Singapore

Library of Congress Catalog Card Number 96-71466
ISBN 0-7651-9425-2

Author: Laura Brooks

Publisher: Robert M. Tod
Editorial Director: Elizabeth Loonan
Senior Editor: Cynthia Sternau
Project Editor: Ann Kirby
Photo Editor: Edward Douglas
Picture Researchers: Heather Weigel, Laura Wyss
Production Coordinator: Annie Kaufmann
Design: Jacquerie Productions

Picture Credits

Arcaid
Richard Bryant 61
Charles Braswell, Jr. 70 (bottom), 72–73, 77, 78–79, 80 (top)
Bullaty Lomeo 28
Sonja Bullaty 34, 117
Butler Institute of American Art 108 (top)
Colonial Williamsburg Foundation, Abby Aldrich Rockefeller Folk Art Center, Williamsburg, VA 87
Dembinsky Photo Associates
John S. Botkin 85 (bottom)
Dan Dempster 106
Darrell Gulin 123
G. Alan Nelson 111, 124–125
Joseph A. DiChello 14
Dick Dietrich 3, 8, 40–41, 104–105, 114, 115
Esto Photographics 6 (top), 14 (top), 30 (bottom), 38 (top), 52, 60, 68 (left), 70 (top), 74, 76, 90, 94, 108 (bottom), 118 (bottom), 126
Grandma Moses Paintings © Copyright 1997, Grandma Moses Properties Co., New York, 42 (bottom), 46 (bottom)
Thomas Hallstein/Outsight 109
Irvin G. Hoover 22 (bottom)
Warren Kimble 7, 62–63
Balthazar Korab 19, 119
Lucinda Lambton 86 (top)
Patti McConville and Les Sumner 45, 58 (bottom), 122
Joe McDonald 15
Steve Musgrove 91, 101
National Gallery of Art, Washington, DC, Gift of Mr. and Mrs. Irwin Strasburger 1
New England Stock Photo 44
Kindra Clineff 48–49
Andre Jenny 84
Johnson's Photography 56–57
Gail Maloney 59
Thomas H. Mitchell 66
J. Schwabel 24–25, 50
H. Schmeiser 31 (bottom)
Paul R. Turnbull 53, 55
New York Public Library 75
John Oram 100
Photolink
Jon Reis 23 (bottom)
Richard Welch 29
Photo/Nats
Gay Bumgarner 118 (top)
Priscilla Connell 65, 110
Michael R. Goodman 43
Photo Network
Faith Bowlus 64
Bill Terry 98–99
Photographers Aspen/David Hiser 112–113
The Picture Cube
Walter Bibikow 69
Claudia Dhimitri 126
Jim Scourletis 54
H. Armstrong Roberts
J. Irwin 127
David Sanger 14, 116
LeRoy G. Schultz 17,18 (bottom), 23 (top), 31 (top), 93 (bottom)
Tom Stack & Associates
John Shaw 102, 103
Stock Montage 30 (top), 51 (bottom)
Superstock 21, 26 (top & bottom), 39, 71
Tom Till 92, 107, 120–121
Unicorn Stock Photo
J. Bisley 9
Jean Higgins 80 (bottom), 88–89, 95
G. Schmeiser 38 (bottom)
Brian Vanden Brink 6 (bottom), 47
Jonathan Wallen 81, 82, 96–97
Terry Wild 16, 32 (bottom), 33, 35, 37 (bottom)
Alan Wycheck 20, 27

Contents

Introduction

Ask any farmer. Barns are supremely practical structures. They come in an infinite variety of colors, shapes, and sizes. They accommodate a menagerie of animals, a tangle of farm equipment, and a bounty of crops. Whether used for storing grain, corralling livestock, stacking hay, or threshing wheat, barns are the heart of the working farm and have endured as the center of American rural life for nearly four centuries.

The fanciful roof lines of this barn in coastal Massachusetts are crowned by a weather vane in the form of a horse.

As the sun rises each morning, the farm awakens in a bustle of activity. For anyone who has ever heard the comforting sound of cows exhaling with warm breath into the crisp dawn air, the first clucks of chickens in the coop, and the shuffle of the farmer's boots on the creaking floor as dawn filters through the wooden slats of the barn, the sensations are forever etched into memory. As if roused by an internal clock, farm creatures gather around the barn to face the new day. The distinctive sounds, sights, and smells of the farm signal the beginning of a

Kissing Cows
by Warren Kimble

enduring ritual that has defined human existence since the dawn of our species. The American barn is the epicenter from which this farm life radiates with each rising sun.

Farming, and therefore barns, have always connoted humble virtuosity. In short, the farmer puts in an honest day's work and returns to the land whatever is taken from it. No one will deny the hard reality of daily toil that farmers face, but people have always had a fundamental respect for the integrity of working the land in harmony with nature. As early as 1769, Benjamin Franklin extolled the virtues of farming and its positive impact on the whole country:

> There seem to be but three ways for a nation to acquire wealth. The first is by *war,* as the Romans did, in plundering their conquered neighbors. This is *robbery*. The second by *commerce,* which is generally *cheating*. The third by *agriculture,* the only *honest way,* wherein man receives a real increase of the seed

At the home of the legendary Buffalo Bill in North Platte, Nebraska, this candy-striped barn greets visitors. Even the three cupolas share the striped motif.

Let me be no assistant for a state,
But keep a farm, and carters.

—William Shakespeare

thrown into the ground, in a kind of continual miracle, wrought by the hand of God in his favor, as a reward for his innocent life and his virtuous industry.

Like farmers, barns seem to transcend their purely utilitarian context. Barns, too, convey a sense of integrity, a clarity of purpose, and a perfect marriage of form and function, of materials and proportion. Standing alone in the landscape, the barn commands as much respect as any public monument across America. There is something about these majestic barns that just seems inherently right. Their materials and their structure appear like a natural outgrowth of the landscape they occupy.

Is there anything so solemn as a lone barn? Barns dominate country pastures like quiet behemoths. Many writers have described barns as spiritual spaces, that they are like cathedrals or country churches in their awe-inspiring atmosphere and mysterious lighting. In these vast, unadorned spaces, there is a quiet monumentality that makes a barn seem an inherently virtuous place. In fact, historically in times of religious persecution, both in Europe and America, people have often retreated to their barns to seek solace and to meet to practice their faith.

REGIONAL VARIATIONS

When I was five years old, my father built a barn on our farm in rural Georgia. He had never built his own barn, and before he began he spent some time driving around, looking at other people's barns, talking to farmers, and making sketches. When he finally settled on a design, he got to work right away. He salvaged an old tin roof from a barn that had fallen to the ground in decay in a neighboring county many years before. He bought new lumber at a local lumberyard, and had the folks at the hardware store mix some paint that would match our house. My grandfather and two neighbors made up the construction crew.

Our barn was a unique creation, but it shared many features and a similar spirit with some of the barns of our neighbors. Even though our barn was constructed relatively recently, within my own lifetime, my father was continuing an age-old tradition of barn building that has persisted for centuries. He looked to his immediate surroundings for models and ideas of what a barn should look like, then incorporated them into a design that suited his own needs.

Like my father, the ideas of the early American farmers about what a barn should look like were influenced by the standing examples in the regions where they lived. When it was time to build a barn, which often occurred even before they finished building their own homes, early farmers looked around their own neighborhood for models. The farmer might take one idea from one neighbor's barn, a second idea from another neighbor's barn, and so on, building a singular creation that nonetheless carried on an architectural dialogue with other buildings in the area. Ultimately, these ideas trace their roots to the regions of Europe from which the farmers' ancestors came.

What is the reason for categorizing barns into regional groups? Early American architecture in general, and barn design in particular, was influenced by the groups of people who settled various areas. Farmers living in different regions drew from different traditions, and adapted their structures to their new environment. At first glance, each barn across the country looks like a unique product, with appendages and additions on all sides. But underneath all the extras, barns can be grouped into certain regional variations. While there are always exceptions to the rules, some characteristics always define regional style. A rich and colorful history accounts for the refreshing stylistic and functional diversity of these barns.

TRADITION AND INNOVATION

When early settlers made their way to the New World from Europe, they brought with them ideas about how a barn should appear and what materials and techniques it took to make one. These preconceived notions were based on what they learned and saw in their own homelands. It is not surprising, then, that the earliest barns in America were based on these European prototypes.

However, these brave settlers were also faced with a new climate, new building materials, and new agricultural challenges in the New World that required a complete reassessment of their old practices and traditions from back home. Therefore, the history of the American barn is a history of adaptation. Early Americans maintained many of the features they admired in their Old World barns, but ingeniously invented new features in response to their new surroundings.

The English settled primarily in the Northeast and the South, and the Swiss-Germans anchored themselves in Pennsylvania, the Mid-Atlantic, and the hinterland of the Appalachians. These pioneers brought a whole host of building traditions with them from Europe, and from these Old World roots sprung innovative approaches to New World barn building. As the early settlers began to fan out across the continent and discover new frontiers, they adapted their barns to the different climates and agricultural conditions they confronted in each region.

The Puritan simplicity of this round barn, painted in traditional red, creates a bold contrast with the rolling,dandelion-covered fields of this Vermont landscape.

The ancient timbers and sagging roof of this California barn are scenic elements of the landscape, but remind us of the fragile condition of many historic barns across the country.

As one who long in populous city pent,
where houses thick and sewers
annoy the air,
Forth issuing on a summer's morn to breathe
Among the pleasant villages and farms
Adjoin'd from each thing met
conceives delight.

—*John Milton*

To the early settlers, the American continent must have seemed like endless miles of forest. Trees were so abundant that it would have been inconceivable to build with any other material. So, in contrast to the stone or wattle-and-daub constructions of the barns they remembered from Europe, timber frame construction became the norm in America. Wrought iron hardware and slate tile roofs were also forsaken in favor of wood; hardy trees were so plentiful, in fact, that wood was used to fashion shingles and even nails and door handles for barns. Wrought iron and slate wouldn't return until much later in America in the form of roofing and door hardware in barns across the country.

FROM SEA TO SHINING SEA

A barn is one of the largest single investments a farmer will ever make. It must be efficient and sanitary, and hardy in its construction to maximize its longevity. It is the heart and soul of the farm economy and an inherent part of the farmer's livelihood. Farmers think long and hard about their needs and finances before building one.

Barns present an intriguing paradox. They are a symbol of stability, endurance, and hard work, but also of the lofty ideals of humans working in concert with nature. Architecturally, these are straightforward, simple, utilitarian buildings, and yet at the same time many barns could hold their own alongside the nation's most noble churches, town halls, or historic monuments. In this sense, barns are indeed American landmarks.

CHAPTER ONE

The Mid-Atlantic

When Europeans visited the New World at the end of the 1700s, the grandeur of Pennsylvania barns astonished them. Though the Europeans were no strangers to grand barns, they were struck by the cavernous wooden giants that graced the new American landscape. They were impressed by the solid, stalwart construction of the Pennsylvania barn, and in this rugged building they saw the robust character of the Pennsylvania farmers themselves. "Farmers in Pennsylvania have a commendable spirit for building good barns," said John Beale Bordley, who visited Pennsylvania in 1799. Arriving in 1789, the Englishman Thomas Anburey exclaimed that Pennsylvanians constructed their barns more solidly than their own houses!

Most barns are architectural agglomerations, including the main barn structure and outbuildings that grow up around it, as in this barn complex in upstate New York.

These European visitors to Pennsylvania were not exaggerating. Many of the early settlers built barns before they built their own homes in this fertile region that today includes such picturesque locales as Bucks and Lancaster

A smoldering sunrise is the backdrop for this working farm in Pennsylvania's Amish country. The silo and windmill became widely used in American farming toward the end of the nineteenth century.

When tillage begins, other arts follow. The farmers therefore are the founders of human civilization.

—Daniel Webster

Counties in the south central part of the state. Since the farm was the source of their livelihood, and the barn was the heart of the farm, farmers wasted no time in erecting these structures, but they also took no shortcuts in constructing these solid buildings—they were meant to last.

THE PENNSYLVANIA DUTCH

The group of settlers who became known as the "Pennsylvania Dutch" actually came from several different European countries. What they shared in common was the German language—most immigrated from Germany, Switzerland, Austria, and Alsace-Lorraine—and their desire to escape religious persecution. These refugees sought religious freedom in the fertile regions of southeastern Pennsylvania over the course of the seventeenth, eighteenth, and nineteenth centuries. Some also settled into neighboring areas such as Maryland, Delaware, New York, and New Jersey.

OPPOSITE
A round barn stands majestically against the blanket of snow that covers this wintry Pennsylvania landscape.

The Pennsylvania Dutch enjoyed a better climate than the English settlers who landed in New England, and were among the most agriculturally prosperous of the settlers to the New World. While their neighbors in New England were trying their luck with tobacco, Pennsylvanians grew wheat as their primary crop. Their barns reflected this prosperity in their large scale and decoration. They also reflected the different building traditions of the national groups that came to Pennsylvania and shared ideas with one another about barns.

Among this diverse group of settlers were the Amish, a religious sect that still inhabits the region today and preserves its traditional way of life. The early Amish often used their barns for religious services; surely they recognized the inherently spiritual quality of the buildings. Amish

Woodwork with fancy geometric patterns grace the gable end of this Pennsylvania barn, an eloquent example of Victorian design.

Barn Decoration

Throughout American history, enormous barns have provided a perfect vehicle for artists, who envisioned on the sides of the barn brightly painted hex signs, fanciful carvings, engaging images, popular sayings, bible verses, graffiti, dates, and more. Barns were only painted after the 1700s in America. Usually painted red to protect the wood, barns were embellished with striking hex signs in the Pennsylvania Dutch country, and with more subdued designs in New England, including star and other geometric patterns.

Today, artists paint fascinating *trompe l'oeil* (a French term meaning "to trick the eye") paintings on the sides of barns, providing the illusion of a landscape or a farming scene; sometimes "fake" windows are painted on the exterior of the building to resemble real ones. Pictures of horses, cows, pigs, and farmers are also common. Some barns even boast copies of the works of the great masters—the Mona Lisa, a portrait of Rembrandt, or scenes from Norman Rockwell's quaint artistic repertory. An often photographed barn in Minnesota is embellished with a purely American red, white, and blue scheme, including a federal eagle on one of the gable ends, and a silo that resembles a barber's pole in its swirling red and blue patterns.

Sometimes words take the place of pictures in barn decoration. In addition to tobacco and medicinal ads of the nineteenth century, occasionally religious messages are painted on the sides of barns. As it does for the advertisements, the barn provides a vehicle for effectively disseminating a moral message to a wide audience. Biblical verses and Christian sayings such as "Teach Me Thy Way O Lord," or "Thou Shalt Not Covet Thy Neighbor's Wife" are particularly common on the historic barns of the Mid-Atlantic states.

But paint is not the only medium for barn decoration. Sculpture and carved wood also embellish barns, from folksy, vernacular carvings of barnyard animals to elaborate, elegant ornamentation that the Victorians so loved, even on their barns. In Pennsylvania, cookie-cutter designs made from large pieces of wood brighten the exterior of many a barn, appearing in wheel and star shapes pierced with lace-like openings. If you look carefully, you might find bales of wheat carved into the keystones of great stone barns in the Mid-Atlantic and New England. Dates are one of the most common things you'll find carved on American barns; they provide an important historical record, and evoke the aura of a by-gone era for modern viewers.

Among the more unusual carved designs on American barns, those cut directly into the side of the barn itself to allow light to filter into the barn through the carved spaces are captivating. From the outside, these designs are not always as striking as painted or sculpted decoration. From inside the barn, however, light filtering through the carved spaces creates a wondrous illuminated pattern within an otherwise shadowy space. These patterns, which include pitchforks, words, chickens, geometric designs, or other images, appear like stained glass windows in a dark church, providing an almost mystical aura inside the building with their play of light. In addition to woodwork, these designs lend themselves to fancy brickwork that graces many farms of the Mid-Atlantic.

On this Pennsylvania barn, the spiky, decorative details of these cupolas serve as a foil for lightening rods.

barns are sometimes the grandest of all in their severe austerity, with unadorned exteriors and interiors that are almost monastic, celebrating the integrity and monumentality of raw materials.

The farmers of the Mid-Atlantic and their barns are the subject of rich folklore and many a local legend. For example, there's the famous farmer who, after learning that rats had infested his barn, burned the building to the ground in order to kill them. This story captured the imagination of the people in the region, and it was used as a metaphor with a meaning not unlike "cutting off your nose to spite your face." In the 1840s, members of a faction of the New York State Democratic party who wanted to abolish corporations in order to get rid of abuses were nicknamed "barnburners."

BUILDING THE PENNSYLVANIA BARN

Until the eighteenth century, farmers in Pennsylvania and the other Mid-Atlantic states generally constructed log barns. Once builders converted to timber framing, they began to build barns on a massive scale, often up to 100 feet (30.5 m) long by 50 or 60 feet (15 to 18 m) wide. In these early days, visitors to the region were amazed at the contrast between these enormous timber barns and the farmers' homes, which remained modest log constructions. In a 1753 visit to Pennsylvania, Lewis Evans remarked, "It is pretty to behold our backsettlements where the barns are as large as palaces, while the owners live in log huts; a sign of thrifty farming." It was also a sign that the barn was the anchor that held together the entire homestead.

Available wood in the region included chestnut, tulip, and poplar, but oak was by far the most popular building material for barns. Its impressive strength made it possible for Pennsylvania barns to span very wide spaces. Massive beams fashioned into gabled frames held the barn erect; for anyone who likes to look at architecture, it is these exposed, time-worn timbers that give the barn its rustic quality and its impressive structural integrity. Wood was so plentiful in the Mid-Atlantic that it was also used to construct floors; only rarely were floors left bare.

From the beginning, many Pennsylvania barns were intentionally built into the side of a hill. While this arrange-

Summer Fruit by Currier & Ives.

Barn Raising

In traditional American communities where it was rare for farmers to carry insurance, a barn destroyed by fire meant much more than simply a ruined building. In many cases, it meant the loss of livelihood for the family. In the face of such a personal tragedy, communal spirit bounded people together in the name of coming to the aid of those in need. The barn raising tradition began.

Community barn raising still occurs among the Amish and Mennonites, and in other traditional American societies. First, the barn raisers construct the wooden frames that serve as the structural skeleton of the new barn. Each separate bay unit is perfectly assembled flat on the ground. When the word is given, each piece is "raised," usually with the help of fifty or more people. What a majestic sight it is as the frame starts to take shape and assume a three-dimensional character, towering over the crowd below. In these traditional communities, it is usually the man's role to construct the barn, while women are responsible for preparing a communal meal consumed once the barn is complete.

The feast that ensues is celebratory, and neighbors joyously break bread together in a festival-like atmosphere. It is a chance for the group to come together to celebrate their accomplishments, their way of life, and their mutual dependence. A barn raising means much more than just erecting a building. It is the locus for bringing a community together for the common good, a time for hard work, and a time to celebrate the communal spirit. Barn raisings are a traditional, community-based effort where everybody plays a significant role.

But barn raisings are not all fun and games. A Pennsylvania tombstone of 1801 reads:

John Moody, 1801
Killed at noon on the fourth of November,
in raising his barn he was hit by a timber.
Be ye also ready for in such an hour
cometh the Son of Man.

In the Amish country, traditional barn raisings are important cultural events that bond the community together in the spirit of hard work and mutual dependence.

ment may seem odd at first, it had several distinct advantages. First, its position, nestled into a hillside, provided insulation and protection from the elements and driving winds. It also allowed easy access to the upper levels from the higher part of the land, and in some cases a ramp led directly into the floor on the second level. Hay and other grains could be stored on the second level, while livestock occupied the space below. These barns are often called bank barns.

The "bank barn," developed in Pennsylvania, stands with one side against an embankment. Haywagons drive directly into the upper story of the barn, while livestock are housed below on the downhill side.

Some Pennsylvania barns are built entirely of stone, and they resemble their Old World ancestors in this regard. One can easily imagine the early Pennsylvania farmer thinking back to the great medieval stone barns he might have seen as a child in Europe while building a new barn on this continent. In formulating ideas for a new barn in the New World, the farmer might have drawn on these ideas and folk traditions that passed through the centuries from Europe to America. Needless to say, these stone barns were built to last.

Classic nineteenth-century lithographs by Currier & Ives celebrated the American homestead during each of the four seasons. Always present in these depictions of American country homes was the nearby working barn.

More often, however, the short ends of the barn would be constructed of stone, while the long sides were built of wood. The wooden sides were of clapboard construction over studs, a sturdy alternative that sheds water and lasts a long time. The entire structure was usually covered with cedar shingles and gabled. A long, sloping roof protected the structure and the animals inside from potentially damaging winds.

Legend has it that an Amish farmer was asked why he went to the trouble to make the stone walls of his new barn nearly five feet (1.52 m) wide. His deadpan reply was, "Why not?" One of the most impressive features of the New World Dutch barn is the thickness of its walls. Whether built of stone or wood, even barns that have survived for over two hundred years leave little doubt that they could survive at least two hundred more.

THE NEW WORLD DUTCH BARN

The barns that were formulated in Pennsylvania and the other Mid-Atlantic states in the 1600s and 1700s charted a course for the future the American barn. The "New Word Dutch barn," along with the "New England barn" that was being developed simultaneously to the north, were the two prototypes that still affect barn design today. The New World Dutch barn was common in Pennsylvania, New York, New Jersey, and neighboring regions.

What are the characteristics of this so-called New World Dutch barn? Its ground plan consists of a central nave flanked by aisles on each side. If this arrangement reminds you of a church, you won't be surprised to learn that it is often called a "basilica plan." This ground plan was well-known in Europe, both in barns and churches, as well as in other types of structures. The basilica plan is of great antiquity, dating back to the final years of the

With its gothic windows and steeple-like cupola, this barn in central Pennsylvania could easily pass for a country church, but the stable doors on the ground floor reveal its true function. The great decorative star on the gable ventilates the interior of the barn.

Roman empire. Thus, the early American barns follow an ancient, tried-and-true building tradition with both functional and symbolic significance.

But the barns erected in the Mid-Atlantic were much larger than the European barns that preceded them; these new American barns are nothing short of cavernous. They also have a plan that is nearly square in contrast to the elongated rectangular shape of European barns. The Europeans almost never used barns to house livestock; barns were primarily for threshing and storing wheat. The New World homesteaders, however, brought their livestock inside the barn to protect them from the harsh conditions of pioneer life in the Americas. Though on the same latitude with their home countries, the climate was much colder on the east coast of the American continent, so farmers had to adjust their farming practices accordingly.

In this New World Dutch barn, animals are relegated to the aisles, while the central nave of the basilica plan is used for wagon storage, and above all, for threshing wheat, one of the most enduring tasks of agriculture throughout the centuries. In many cases, this barn has two stories, with the threshing floor on the upper level, and livestock below. In "bank barns," farmers access the upper level via a ramp on the bank of the upward slope of the hill; they can drive a wagon directly into the upper floor of the barn from ground level.

Modern-day visitors to the region often comment on the overhangs that project from the second story of the barn over the first story. These "overshoots" are one of the most distinctive features of the Pennsylvania barn. This cantilevered design, sometimes called a laube, protects the livestock that congregate under its shelter. It can also be used to ensure that bales of hay or other crops stay dry while they await storage

A corn crib stores feed for the winter behind a modern barn. Although industrial techniques have revolutionized farming in the last fifty years, the barn remains the heart of traditional farming tasks that have defined the rural way of life for centuries.

Separating the Grain from the Chaff

The word "barn" derives from the Old English words *bere* and *aern,* which, roughly translated, mean "barley house." These Anglo-Saxon roots provide a clue about the original function of the barn in the Old World.

In Europe, barns were first and foremost a space to thresh wheat, or to separate the grain from the worthless chaff of a newly harvested crop. Great bales of wheat were carried in from the fields and loaded onto the threshing floor through the giant doors on either side into the heart of the barn. Farmers used hand-held tools called flails to dislodge the empty husks from the precious grains of wheat. Alternatively, the wheat could be laid out on the threshing floor, and horses would be led to walk over it in order to accomplish the same goal. Medieval manuscripts depict people threshing wheat; it is an ancient practice that has affected barn design for centuries.

In Colonial America, farmers carried on the traditions of their European forefathers, and early barns centered around a giant threshing floor. Today, farm machines called combines do the work that threshing used to accomplish, and the great threshing floors of historic barns are now used for storage or to accommodate expanding livestock stalls. Ever-changing needs in American farming transformed the barn from solely a threshing space to the truly multipurpose structure that makes the American barn a truly unique creation.

By the way, threshing wheat is preserved in our language if not in current farm practice. Threshing took place in the central threshing floor with the two doors on either end open, so that the wind could carry away the lightweight chaff, leaving the grain behind. In order to keep the valuable grain from blowing away, farmers inserted boards into slotted spaces in the bottom of the doorway. Boards wedged between the door jambs thus came to be known as—you guessed it—"thresholds."

The barn door on sliding tracks has numerous advantages over the traditional hinged barn door. It takes up less space, resists blowing in the wind, and is a cinch to open and close.

As this nineteenth-century engraving details, the barnyard has always been a bustle of activity, alive with movement and noise.

inside the barn. This overshoot may derive ultimately from European precedents. Many houses in Europe, particularly those of the mountainous, German-speaking regions from which these settlers came, incorporate a similar construction into their façades.

Barn builders often made large double doors on the short ends of the barn so that wagons could drive in one side and out the other in order to load or unload grain. These doors often measured up to 10 or 15 feet (3.05 to

4.57 m) wide. In 1748, the Swedish traveler Peter Kalm marveled at this type of barn on a visit to New Jersey: ". . . in both ends of the building were large doors, so that one could drive in with a cart and horses through one of them, and go out at the other." This arrangement also provided much needed ventilation and light, since Mid-Atlantic barns were generally without windows prior to the nineteenth century. Later, ingenious barn builders invented sliding wagon doors on tracks that were less cumbersome to open and took up much less space.

In Mid-Atlantic barns, the roof lines are very steep, with sharp, dramatic gables and low eaves over the aisles. After the middle of the nineteenth century, gambrel roofs became popular in order to increase space in the upper story. The distinctive gambrel shape that most Americans recognize as a distinctive trademark of a barn has two slopes, and therefore opens up the space over the eaves of the barn,

Oh, farmers,
pray that your
summers
be wet and your
winters clear.

—Virgil

The gambrel roof, the classic roof type for American barns, distinguishes many barns of the Mid-Atlantic states. In this barn and outbuildings, the triple gambrel roofs are particularly striking.

Why Are Barns Red?

Legend has it that a Pennsylvania Dutchman was once asked what the proper color of a barn was, to which he replied, "Any color is a good color as long as it's red." Why is it that many American barns are red? Through the 1700s, most barns were left unpainted, but farmers eventually wanted to protect the wood of their barns to maximize their longevity.

The substance that gives barns their red color is actually a wood protector made from a mixture of iron oxide, available in the soil, along with skim milk, linseed oil, and lime. Some early farmers even claimed that their secret ingredient was turkey blood, which enhanced for the dark shade of red. Farmers mixed these ingredients into a paint that hardened into a vibrant hue. The iron oxide made the coating durable and plastic-like, protecting the barn from all elements.

This dark red color absorbs the winter sun's rays, keeping the building warm. Therefore, red barns are more common in northern states, where warmth is paramount in the face of harsh winters. In fact, red barns are also common in Scandinavia, where they are particularly striking against the perpetual snow. Conversely, southern farmers generally have turned to lighter colors that reflect the sun, maintaining a cooler temperature on the inside of the barn. Even in Pennsylvania, white and yellow recently have become more common than the traditional red. Rustic, naturally weathered barns are also common, especially in New England and the South.

The Mystery of Hex Signs

Visitors to southeastern Pennsylvania are struck by the fanciful, brightly painted forms that decorate the sides of barns and sometimes silos in that region. These so-called "hex signs" have long been the subject of controversy. What do they mean? Was the hex sign meant to ward off evil or to protect livestock from witches, as some have suggested? After all, the German word for witch is "hex," a term which plausibly might have been imported along with German immigrants to the region, and this belief may derive from ancient superstitions left over from the Old World. Or, as others believe, are hex signs nothing more than striking examples of folk art, meant to tease the eye and delight the imagination?

Whether or not the hex sign effectively wards off evil, many of the specific forms that hex forms take are steeped in folklore. If you want to be sure that lightning does not strike your barn, so the saying goes, paint a six-pointed star on the outside of the barn. If you want to keep the devil away from your cows, paint a white outline around the barn doors. Better yet, dupe the preying demons altogether by painting false doors on the exterior of the barn, in white paint, of course.

These motifs are most common in ten counties in southeastern Pennsylvania. Traditionally, hex signs are based on the circle, usually with geometric shapes, star patterns, swirls, or other radial motifs painted in bright, primary colors. Eight-pointed stars within circles, rosettes, and wheel forms dominate these designs. Plain, solidly painted stars with no circular border, called *schtanna* in the local dialect, are also common.

Against the backdrop of a dramatic red barn, these symbols are potent reminders of our forefathers and the objects of fascination. But these designs are not limited to barns. Similar forms also appear on historic gravemarkers in the region. They also show up on folk furniture, from chairs to chests and baby cribs. Some historians see Christian symbolism in these unusual, kaleidoscopic hex signs that grace so many barns. In fact, some of these forms are also found on churches in the region, especially the common star-shaped ventilators that pierce the gables below their belfries.

The earliest accounts of hex signs come only after the Civil War, in the 1860s. Since the 1940s, hex signs have been one of the biggest tourist attractions in southeastern Pennsylvania. In fact, shops selling hex signs on wooden disks abound. European visitors, especially from the highlands of Switzerland, southern Germany, and the Basque region of the Pyrenees, find great affinities between Pennsylvania hex signs and forms carved on architecture and furniture in these regions of Europe. The six-pointed star, for example, was carved for centuries on cabinets and mantelpieces in Europe. Do these forms have an ancient history steeped in the traditions of the Old World? The origins of these fascinating forms remain a mystery.

allowing for a spacious hayloft. The word gambrel comes from the hock or hind quarter of a horse, which is similar in shape.

Once developed, the New World Dutch barn was an impressive affair. In 1882, Gertrude Lefferts Vanderbilt evoked its mysterious, distinctive qualities:

> The barns of the Dutch farmers were broad and capacious. The roof, like that on their houses, was very heavy, and sloped to within eight or ten feet of the ground. There were holes near the roof for the barn swallows that flitted in and out. . . . Through the chinks of broken shingles the rays of the sun fell across the darkness as if to winnow the dust through the long shafts of light, or where the crevice was on the shady side, the daylight glittered through like stars, for there were no windows in these barns; there was light sufficient when the great double doors, large enough to admit a load of hay, were open.

BARN DECORATION

Pennsylvania barns are perhaps most famous for the so-called "hex signs" that decorate their exteriors. Against the red backdrop of the barn, these colorful and mysterious signs lure and intrigue the visitor. Painted by hex sign artists or by farmers themselves, hex signs usually consist of a circular motif with geometrical, often kaleidoscopic designs in green, white, red, and black. The circles can be up to 5 feet (1.52 m) in diameter, and some barns have as many as eight or

Star-shaped hex signs make curious patterns against this red barn. Some believe these hex signs are meant to ward off evil spirits, while others claim that they are simple expressions of a local folk aesthetic.

twelve designs decorating one of its sides. Many patterns were repeated over and over in different places, perhaps conveyed through pattern books or a standard repertory of traveling artists.

Another popular form of decoration for Mid-Atlantic barns are the brick patterns that, when seen from the interior of the barn, illuminate the dark interior with distinctive patterns in the shape of stars, horses, people, or geometric designs like diamonds and hearts. These designs showcase the true art of the bricklayer and the ingenuity of the early American artist.

Into my heart and air that kills
From yon far country blows:
What are those blue remembered hills,
What spires, what farms are those?

—A. E. Houseman

An idyllic field of flowers serves as a carpet for this barn in Pennsylvania's Amish country.

CHAPTER TWO

New England

According to legend, one bleak February day in the 1700s, a Maine farmer began making his way from his house to the barn to tend his livestock. Suddenly, a violent snowstorm erupted, and the unfortunate farmer was found the next day face down in the snow—he never made it as far as the barn.

"If you don't like the weather, wait a minute," goes the well-known adage in New England. Predicting the ever changing weather is paramount for the New England farmer, and it stands to reason that some of the earliest American weather vanes developed in the Yankee states. It is also not surprising, given stories like that of the Maine farmer, that New Englanders were instrumental in developing the connected barn, in which the barn is attached to the dwelling house, avoiding the need to go outside at all.

Weather always determines what form a barn will take. When brave settlers first set foot on New England shores in the 1600s, they faced winters the likes of

This standing Native American figure appeals for rain. The weather vane is an invaluable tool for the savvy farmer, who relies on it to plan the annual cycle of planting and harvest.

Currier & Ives' *Winter in the Country—Cold Morning, New England* captures a busy farm on a frosty morning.

I reverently believe that the Maker who made us all makes everything in New England but the weather.

—Mark Twain

A brilliant red saltbox barn compliments the vibrant fall foliage in Vermont.

Barn Dance

The old nursery rhyme that begins, "Old MacDonald had a farm" has us believe that farms are pretty noisy places to begin with. But a cluck-cluck here and a moo-moo there is nothing compared to the hullabaloo that ensues when a barn dance invades the barnyard. A barn is a natural place for a hoe-down; all you need is some fast-stepping music and a few good neighbors. It's the only space around that's big enough to accommodate a large group of fiddlers, square dancers, freckle-faced adolescents, and clapping and foot-stomping old folks all at the same time.

Traditional throughout rural America, the sound of fancy fiddling drifting across the hills drew people from all over in mule-drawn wagons to the barn dance. New England farmers took advantage of warmer weather to socialize, sing, and dance, as captured in Grandma Moses' *The Barn Dance*. These country socials were equally popular in the rural sounth—in fact, Grand Ole Opry began in 1925 under the name *WSM Barn Dance,* and maintained the flavor of a barn dance with its country music favorites.

Square dancing and clogging make barn dance music come alive. Dancers in gingham dresses, overalls, string ties, straw hats, and red bandannas fill the floor with color as the clickety-clack of tap shoes resounds in the cavernous barn. These traditional folk dances originated in England, Scotland, and Ireland, and were brought to the New World along with early settlers, where they evolved into their current form. The "caller" is a vital component in square dancing and clogging. The caller's voice echoes through the barn, calling "sets" or instructions for certain dance steps like "birdie in a cage," "dive for the pearl," "swing your partner round and round," and "all promenade."

Primarily places for hard work, barns also hosted some of rural America's biggest social events. Grandma Moses' 1950 painting *The Barn Dance* depicts one such gathering, complete with musicians, frolicking young couples, and food and drink.
Oil on canvas, 35 x 45 in., K920.

which they had never seen in Northern Europe. Like their earliest homes in the new land, the barns of the New England settlers were built close to the ground or against a slope; they were compact, strong, windowless, practical, and Spartan.

ECHOES OF THE OLD WORLD

Although the earliest English settlers to the northeast forsook their homeland to escape religious persecution in the New World, these settlers nonetheless retained the ideas about barn types and the special barn-building techniques they had learned in England. The early barns of New England were based on English prototypes.

The salient characteristic of the English barn is its division into three units called bays. The threshing floor occupies the central bay, and the two flanking bays, which are used for storage, are called mows. Like the New World Dutch barn, the English barn incorporated large doors on either side so that wagons could drive in one side and out the other. But in contrast to their Pennsylvania neighbors, the New England barn

The gable roof is the oldest and most basic roof design for barns. Later, innovative barn builders sought other solutions like the gambrel and gothic roofs, which allowed for more space in the upper story.

A dairy barn in Vermont incorporates a low, sloping roof that guards against harsh New England winds.

OPPOSITE
The classic red color of barns is achieved through a homemade recipe for wood protection that includes lime, linseed oil, and skim milk.

While the cock with lively din
Scatters the rear of darkness thin
And to the stack, or the barn door
Stoutly struts his dames before
Oft list'ning how the hounds and horn
Cheerly rouse the slumb'ring morn

—John Milton

Grandma Moses and the Folk Aesthetic

Quilting bees, country fairs, sleigh rides, snowball fights, holiday traditions—no other artist celebrates the joyous qualities of rural farm life than Grandma Moses. Ever with an optimistic tone, the charming paintings of Grandma Moses capture an idyllic rural aesthetic that centers around the barn and the country folk who inhabit it. Her depiction of barns is historically accurate, and in Grandma Moses' paintings, the barn always serves as the focal point for the happy, harmonious cohabitation of human and animal elements of the farm.

Anna Mary Robertson Moses, later known as Grandma Moses, was born in 1860 and lived to be 101. The daughter of farmers, she and her husband, Thomas Salmon Moses, worked as tenant farmers in Virginia. Grandma Moses also lived in Vermont and upstate New York, and her depiction of farm life is often associated with New England because of the architecture of the houses and barns she paints, and the rolling countryside typical of the region.

In a painting entitled *The Barn Dance* of 1950, the artist depicts a three-bay English-style barn typical of rural New England; the great doors of the barn are open on both sides, framing a picturesque rural landscape beyond. Grandma Moses celebrates the vitality of the barn dance with lively groups of dancers, fiddlers, and accordionists, as well as horse-drawn carriages, and food and drink. In all of Grandma Moses' paintings, everyone is smiling.

Grandma Moses uses the image of the barn to symbolize harvest time. In a painting entitled *Pumpkins,* she depicts farmers threshing wheat in the great threshing floor of a drive-through English-style barn, their flails raised above their heads with concerted effort. Haywagons prepare to deposit bales of hay into the barn.

Honest and direct, it is easy to understand how Grandma Moses' paintings grew out of her experience in sewing, quilting, and embroidering. Her paint strokes resemble stitches in a marvelous quilt, and have the same country, folksy quality that distinguishes her as one of America's most significant folk artists.

In this Grandma Moses painting entitled *Pumpkins*, the artist's characteristic optimism belies the difficult task of threshing wheat. Farmers raise their flails on the threshing floor of this great English-style barn.
Oil on pressed wood, 16 x 24 in., K1380. Copyright ©1996, Grandma Moses Properties, Co., New York

builders placed these doors on the long ends of the barn as opposed to the short ends.

What a fascinating contrast the English barn and the Pennsylvania Dutch barn make. The English barn is smaller than its enormous Pennsylvania cousins, and is sometimes left unpainted and undecorated, its wood curing to a natural, driftwood-like gray. These no-nonsense barns, like the austere tobacco barns of the Connecticut river valleys, contrast sharply with the intricately painted examples from south central Pennsylvania and other states in the Mid-Atlantic region.

This expansive horse barn in Shelburne, Vermont is a striking example of the vast size and elaborate construction of many New England barns.

Their European predecessors rarely housed livestock inside barns, but early settlers to New England quickly enlarged and adapted their barns to protect farm animals from the cold climate they faced in the New World.

FORGING A NEW IDENTITY

Early New Englanders quickly adapted their barns to suit their needs in the New World. In the Old World, the English had built their barns of stone or brick, usually with thatched roofs. Like the Mid-Atlantic farmers, however, settlers to New England immediately capitalized on the plentiful wood in the region. In Colonial New England, forests were so dense that trees grew straight toward the light, making for very straight timbers that lent themselves to vertical framing. Pine and hemlock were used most often for timber framing New England barns.

Their English ancestors usually had roofs of thatch for houses and barns, and New Englanders followed suit. But the harsh winters of New England and the great fire hazard the thatch roofs prompted them to switch to clapboard and shingles to roof their barns. Newly forged legal codes went so far as to prohibit settlers in some areas of Massachusetts to construct thatch roofs altogether.

Europeans usually did not house livestock in their barns; barns were primarily for grain storage, and there were separate facilities for livestock—stables for horses, sties for pigs, and byres for cattle. The English barn was above all for threshing grain. The three-bay barn was well adapted for this task, since the farmer could drive a wagon through the middle of the threshing floor, and store crops on either side. Thus, the English barn was not well suited to house livestock.

Originally, like their European predecessors, New Englanders left their livestock outdoors during the winter and used the barn only for threshing and storage. Over time, however, farmers built separate stables, then incorporated these spaces into the barns themselves. Town permits in Boston record numerous petitions from settlers in the 1700s wishing to construct "cow houses" on their property: "It is permitted John Smith to

The hardy construction of this barn and rock fence in New Hampshire illustrates the tenacity of early New England farmers.

Lofty Ideals

Having a suitable place to protect hay from storms and dampness has always been a major concern of the farmer. Haylofts in European and early American barns were accessible only from the inside, usually via a ladder leading from the main threshing floor to a loft built under the eaves of the barn's roof. Grain and hay continued to be stored alongside the threshing floor, so the hayloft was not a critical element in these early barns.

But as the American barn developed, and livestock and farm equipment began to occupy the lower level of the barn more often, the fully developed hayloft accessed from the outside came into being. Barns are usually three-bay, two-story structures, and once corn and other crops were used to feed livestock, grains were stored in the loft above, and livestock were housed below.

Haylofts in the South often have a peak-like section that protrudes beyond the end of the barn; these are often referred to as hay gables or hay bonnets. Hay bonnets are also found occasionally in New England and the Midwest. Below the hay bonnet, double doors swing out to provide access to the hayloft from the outside. A track installed along the full length of the barn's ceiling and extending out to the hay bonnet allows a pulley to lift a giant hayfork up and down, elevating the hay from the ground into the hayloft.

Giant tines grab bales of hay, hoisting them up and depositing them in the loft. This job requires at least two people, one on the ground hooking the tines of the hayfork to the bales of hay, another in the loft to operate the hayfork and maneuver the placement of the hay bale once it arrives in the upper story.

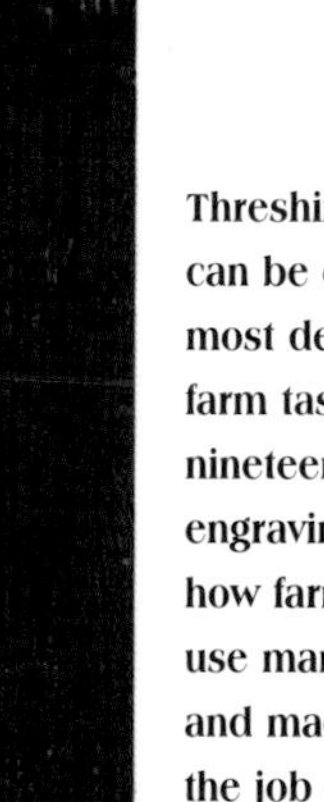

Threshing wheat can be one of the most demanding farm tasks. This nineteenth-century engraving shows how farmers made use man, animal, and machine to get the job done.

American folk artists like Rose Labrie (b. 1916) painted rich farm scenes based on their childhood memories. Labrie's *On a Sunday Afternoon* (1979) captures a piece of rural life on the farmer's traditional day of rest.

A farm is like a man—however great the income, if there is extravagance but little is left.

—Marcus Porcius Cato (Cato the Elder)

construct a cow house on his land on Fish Street," reads one such deed. Animals eventually moved inside the barn for protection from the elements. Since New England winters are harsh, farmers relied on aisles on either side of the threshing floor to accommodate livestock that could not remain outside all the time.

A NEW PROTOTYPE

By creating aisles alongside the threshing floor to accommodate livestock, New Englanders ended up with a barn that had a ground plan similar to the New World Dutch barn. After the middle of the nineteenth century, New Englanders moved the wagon doors from the long sides to the short gable ends of the building, also like their Pennsylvania Dutch neighbors. This new creation, known as the New England or Yankee barn, filtered from New England into Canada and the Midwest over the course of the nineteenth century.

Many barns in New England weather to a natural gray, resulting in an austere and rustic beauty that characterizes the region.

ABOVE
Often barns are constructed of mixed materials. Wooden frames built on sturdy stone foundations like this one are common in New England and the Mid-Atlantic.

OPPOSITE
In New England, low, sloping roofs are critical to keeping the barn and its inhabitants warm. Snow piles on the roof, creating an effective layer of insulation against the harsh climate.

New Englanders initially used simple, steep gable roofs that easily shed snow. Later, like their Mid-Atlantic neighbors, they preferred the gambrel roof since it provided more space under the eaves. More room in the upper story is particularly important when livestock are housed below, since the farmer needs more room to store hay on the second level.

Despite austerity of many New England barns, they often don fanciful cupolas, particularly from the Victorian Age. These picturesque towers with graceful, almost lacy carvings are sometimes as ornate as a wedding cake. Visitors to New England may spy decorative carvings in the gable end above the doors, including wheel-and-spoke patterns, and star shapes, which are particularly popular in New Hampshire.

PRECEDING PAGE
Early visitors from Europe to the New World were amazed at the tremendous size of American barns. Many barns went up before houses, since the barn was the heart of the family's livelihood in Colonial days.

A Crowning Achievement: The Cupola

Cupolas, the tower-like structures that crown many barns across America, spark the creative imagination of American barn builders. Like weather vanes, cupolas play both a functional and decorative role. Cupolas were practically unknown until the mid-nineteenth century, but along with other technological advances in American barn architecture such as silos, sliding doors, and gambrel roofs, they revolutionized the practicality of the American barn and added a new dimension to its visual appeal.

Usually square and pierced by louvered slats on at least two sides, a cupola provides ventilation and filtered light inside the barn while still protecting it from the elements. It allows the heat that collected in the upper story of the barn to escape through the slats. In addition, it often serves as a base for lightening rods and weather vanes.

The cupola is often the preferred spot for whimsical decoration. Although the vast expanses of space in some barns are left unadorned, the cupola is almost always decorated with paint or carved wood. In the Victorian era, artists exploited the decorative possibilities of the cupola by carving lace-like designs and gothic windows reminiscent of church architecture. Following the prevailing Victorian aesthetic of the day, many of these fanciful creations resemble wedding cakes in their frilly details.

Perched atop the cupolas of American barns, weather vanes are both creative and practical. Their motifs, like this tractor, are often drawn from daily life on the farm.

TOBACCO BARNS

The Connecticut river valleys have just the right climatic conditions for growing tobacco. While most of the nation's tobacco is now produced in the South, these austere, undecorated behemoths still grace many fields in the rolling hills of Connecticut. New England tobacco barns are usually undecorated, weathered to a natural gray. They are windowless, so they present a severe, monumental, and sober exterior that dominates the surrounding fields that follow the rolling landscape.

Many tobacco barns use vertical siding; wooden planks are fitted vertically along the frame as opposed to laterally or horizontally. This technique serves a practical function, since these slats can be opened to allow for ventilation, which is paramount in tobacco farming. The slats open slightly from the bottom, or have hinged panels with poles that allow the slats to open easily with a single lifting motion. Air then passes over the tobacco, which hangs from wooden racks on the inside during the long curing process.

THE SALTBOX

The saltbox design, which is used extensively for Colonial homes as well as for barns, is quintessentially New England and is most well adapted to inclement weather. The saltbox has a steep gable, with a long, sloping roof that almost reaches the ground on one side. The long roof is oriented to the direction of the prevailing winds, deflecting the harsh elements and insulating the interior of the barn.

Country journals of the seventeenth and eighteenth centuries explained how to

Ramps offer easy access on the long and short sides of this New Hampshire barn for the haywagon, farmer, and livestock.

OPPOSITE
The circular form of the barn in Hancock Shaker Village ensures an efficient use of space. Livestock are housed around the perimeter of the barn, while hay and grain are threshed in the center.

pile fallen autumn leaves against the low wall, providing a layer of insulation against extreme weather. When the snow fell in the winter, it piled up on the sloping roof, insulating the barn from the cold and protecting the animals inside. An old almanac recommended: "Slope your barn 'gainst northern blast, and heat of day is made to last."

THE ROUND BARN

Though today one or two round barns can be found in nearly every part of the country, and the Midwest now boasts more round barns than any other region, the earliest examples come from New England. The most famous is the perfectly circular barn constructed in 1824 in the Shaker colony of Hancock, Massachusetts. This round stone structure, whose unique architectural clarity, clean lines, and logical organization draw visitors from all over, is a particularly striking example.

BELOW
The famous round barn in Hancock Shaker Village, Massachusetts, has inspired many folk artists to portray its wide girth and impressive stone construction.

©WA

Kissing cows welcome us into this delightful painting of a round barn by Vermont folk artist Warren Kimble.

Though little, I'll work as hard
as a Turk,
If you'll give me employ,
To plow and sow, and reap
and mow,
And be a farmer's boy.

—*Anonymous*

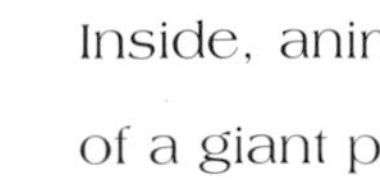

Inside, animal stalls radiate outward from the center pier like wedges of a giant pie. This cavernous barn measures 270 feet (82.35 m) in circumference, 90 feet (27.45 m) across, and can accommodate ten haywagons and 54 cattle. As in the Amish societies, for the Shakers barn building was a communal undertaking. Their societies were experiments in communal ways of life, and the round barn, which is often said to have symbolic properties, fit into their utopian world view. It probably took twenty or more people to lift some of the massive beams and stones that make up this unique barn.

OPPOSITE
Picturesque curved roof lines reminiscent of gothic arches create a sense of musical harmony in this Vermont farmscape.

In addition to its picturesque qualities, the round barn has its practical side. Its shape allows the farmer to maximize interior space while minimizing exterior walls. Hay placed in the center of the circle is accessible to the farm animals in the stalls radiating from the center. The practical Shakers had a clear definition of perfection: "anything may, with strict propriety, be called perfect which perfectly answers the purpose for which it was designed." According to this definition, the Shaker barn at Hancock Village is as about as perfect as a barn gets.

BELOW
While barn doors on rolling tracks became more common toward the end of the nineteenth century, we still find uncomplicated hinged doors like this one, latched with a simple wooden board.

CONNECTED BARNS

Intrigued by the ways of the early New England farmers, an Englishman visiting the region in 1780 reported to his family in England, ". . . these people live near their animals . . . it is difficult to tell where their house ends and their barn begins." In New England, as elsewhere, barns grow appendages and change with the times and the lives of their builders and owners. They can be adapted, added on to, and expanded over time.

You only have to think back to the legend of the unfortunate Maine farmer caught in a blizzard to understand why New Englanders developed the connected barn. In cold climates, barns are often attached to the house by a passage or breezeway, eliminating the need to be exposed to the weather while doing chores. Houses can be attached in an L-shape or sometimes in the same axis as the barn. Many examples are found in Maine and Vermont. Intervening spaces between the dwelling house and the barn might include a kitchen, tool room, milk room, or other utility rooms.

Any New England farmer will tell you that a connected barn also facilitates the constant maintenance required to keep farm buildings in working condition. The yearly cycle of barn maintenance in New England involves repainting and repairing parts of the barn each spring after a harsh winter, and preparing the building to weather the winter cold. As the cows grow shaggy winter coats in anticipation of the cold,

Connected barns are common in New England. On this farm in coastal Massachusetts, the occupants don't have to brave the snow when walking from their farm house to the barn.

Roofs

There is a lot of discussion about barn roofs. And roofs *are* important. A barn, after all, is primarily a shelter, and a roof is the integral part of this equation. In fact, a special type of hay barn prevalent in some parts of the South and the Midwest is nothing more than stilts with a roof to keep out the elements. Farmers have many choices available to them in barn roofs. Different types of roofs accomplish different functions; many are based on historical or regional models common in a certain area.

The oldest type of barn roof is the gable roof. The gable is simply the triangular upper part of the wall that meets the roof of the barn. It can be steeply pitched to shed snow, so this type of roof is common in New England, for example, where winters are harsh. But the gable roof can prove impractical, especially if your barn has a second story. The sharply pitched roof limits space in the upper story, and you're likely to bump your head when working under the eaves.

A gambrel roof alleviates spatial problems on the second floor, although it is a bit harder to construct. Unlike the gable roof, which consists of essentially two sloping planes, the gambrel roof uses four planes set at an angle on each side. No one knows for sure what the origin of the gambrel roof is, but it seems that barn builders developed it in order to provide more space for hay or other materials on the second floor. The gambrel roof appeared in America only after the 1850s, and companies offered pre-fabricated gambrel trusses by the end of the century, greatly facilitating building this tricky roof type. The gambrel roof is ubiquitous today, not only for barns but also for suburban homes and even fast food restaurants advertising down-home fare.

Although the gable and the gambrel roofs are the most common in the United States, some other roof types can be found throughout the country. In the late nineteenth century, picturesque curved gothic roofs appeared, and these continued to be built beyond the Victorian Age well into our own era. The hip roof, which fits snugly on top of four walls of equal height, is also common.

farmers must also prepare their barns to face the harsh elements. These preparations usually include insulating the barn against the frigid winds that sweep the farmlands of the northeast. Despite all these precautions, however, keeping such a large interior space warm is often futile in so harsh a climate. It is no surprise that the popular metaphor "as cold as a barn" originated in New England.

Ah too fortunate farmers, if they knew their own good fortune!

—Virgil

Someone left the great rolling door of this historic barn in Vermont ajar, inviting us to explore its awe-inspiring interior spaces.

CHAPTER THREE

The South

Barns of the South suffer a much maligned status. New England and Mid-Atlantic farmers have historically looked down their noses at the farm structures of their southern neighbors. To northerly farmers with barns on a grandiose scale, the early barns of the South weren't really worthy to be deemed "barns" at all—they were mere sheds. To outside observers, their small size, weathered, unpainted exterior, and interior spaces open to the elements left a lot of room for improvement.

But this disdain is unwarranted. In contrast to the vast plains of the Midwest or the rolling hills of New England, the South is characterized by wooded expanses and rocky soil. The region is not subjected to same climatic rigors that challenge northern and Midwestern farmers. As in New England and the Mid-Atlantic states, southern farmers ingeniously adapted their farming structures to their own unique environment. Consequently, southerners developed several new types of barns specifically designed for their needs.

In the foothills of the Blue Ridge Mountains, this structure typifies the Southern barn with its large, central opening flanked by ventilated bays on either side. The metal roof is common in the South, amplifying the heavy raindrops that fall frequently in the region.

The farmyard went right up to the doorstep of the family home in nineteenth-century America, as this Currier & Ives lithograph depicts.

Only Yankees and fools predict the weather.

—*Popular proverb*

OPPOSITE
The verdant, rolling hills of Tennessee provide a picture-perfect backdrop for this rustic barn. The gambrel roof has become a classic in American barn design.

FOLLOWING PAGE
Unchinked logs and rustic, time-worn timbers give this barn in the mountains of Tennessee an enduring, picturesque quality.

In Louisiana, an antiques buff transformed a large English-style barn into the perfect workshop for restoring cane chairs and other antique furniture. In North Carolina, tobacco barns have been turned into museums celebrating centuries of rural farm life in the region. As in the Amish country, southerners in the eighteenth century exploited the lofty atmosphere of the barn, gathering there for religious services. Traveling ministers used barns when churches were too far away or not yet constructed. The Englishman John Wesley, founder of Methodism, conducted evangelistic preaching in barns as well as in houses and under trees during the 1700s in the rural South.

BUILDING THE SOUTHERN BARN

In the 1600s, the English colonized the southern coast, including the Chesapeake Bay. German-speaking groups also filtered down into the southern hinterland from Pennsylvania. The mixture of these two cul-

This is the farmer sowing the corn,
That kept the cock that crowed in the morn,
That waked the priest all shaven and shorn,
That married the man all tattered and torn,
That kissed the maiden all forlorn,
That milked the cow with the crumpled horn,
That tossed the dog
That worried the cat
That killed the rat
That ate the malt
That lay in the house that Jack built.

—Anonymous Nursery Rhyme

Weather Vanes

"Ring around the moon, brings a storm soon." "Rain before seven, stop before eleven." "When cows lie down it's a sure sign of rain." "The higher the clouds, the fairer the weather." Popular folk sayings like these reflect generations of farming experience, and early farmers undoubtedly relied on their intuition and observation of the conditions of the moon, their cows, and the clock in tending their crops and planning for the harvest. But farmers have another more reliable tool. Swiveling in the breeze, the weather vane points the direction of prevailing winds. For the farmer, the weather vane is invaluable for predicting changes in the weather.

Weather vanes can be traced as far back as ancient Greece. The Vikings also used metal weather vanes decorated with fantastic creatures from Norse myth. Old World farmers fashioned weather vanes based on heraldic motifs and animal forms, particularly that of the rooster.

In America, weather vanes top cupolas in a decorative display of local traditions. Favorite motifs for weather vanes include barnyard animals like the cock, the cow, and the horse, as well as patriotic symbols like the federal eagle. Weather vanes became popular in America around the mid-nineteenth century, made famous by renowned makers like L.W. Cushing and J.W. Fiske. Weather vanes fit the Victorian taste for fancily decorated objects, and were ubiquitous on farms at the turn of the century. Today they are picturesque collectibles, found in antique shops and folk art museums all over the country.

tures and their different traditions of barn building influenced barn construction in the South. The best features of the English and the German/Dutch barn were combined, and in the process a completely new form of barn was adapted to the prevailing climatic and agricultural conditions of the southern states.

The mild climate of the South has always affected barn design in the region. The constantly moderate temperatures of the South are conducive for growing a variety of crops in the southern states. Unlike their northern counterparts, southern farmers have less of a need for architectural features that insulate or protect the barn from the elements; sloping roofs, connected barns, and insulation, for example, are rarely found in southern barns. In fact, many southern barns are open to the outside—some have no doors, and others incorporate lean-to sheds alongside the barn with no walls at all. Only a very temperate climate allows for these adaptations.

Southern soil and weather were less well adapted to growing wheat, so early settlers moving from Pennsylvania south into Virginia

A warm climate allowed for more open, ventilated spaces in early barn architecture of the American South. A metal hipped roof shelters historic haywagons in this North Carolina barn.

OPPOSITE
In this North Carolina tobacco barn, spaces between the logs provide the ventilation necessary to properly cure a crop of tobacco.

The farm was crouched on a bleak hillside, whence its fields, fanged with flints, dropped steeply to the village of Howling a mile away.

—Stella Gibbons

Southern barns often consist of an enclosed central space flanked by two open aisles. These lean-to spaces, covered with a tin roof, offer the perfect place for storing farm equipment and tying horses.

planted corn crops instead. Therefore, there was no longer a need for the vast threshing floor that was the centerpiece of the New England and Pennsylvania barns to the north. In fact, the traditional crops grown in the South—cotton, peanuts, and sugarcane—do not require barns at all. Simple sheds are all that is needed to store these crops after the harvest. The exception, of course, is tobacco, a primary product of the South today that requires large tobacco barns of the type also found in New England for curing and storage.

Because the English first colonized the South, the earliest barns in the region follow the English barn prototype that was common in Colonial New England—a three-bay structure with a central drive-through threshing floor and hay mows on either side. Many of these early barns were log constructions with lean-to additions that anyone who has visited rural Tennessee and Virginia will recognize. In contrast to their northern neighbors, southerners tended to build barns on a smaller scale.

An austere tobacco barn towers over crops on this Kentucky farm. The vertical slats of the barn open slightly to allow for ventilation, but there are no windows.

The Barn Door

A legendary Alabama hog farmer learned a lesson about barn doors the hard way. Constructing a new barn to house hogs, he underestimated the size that these young porkers would reach as adults. The first sow to top 300 pounds found it impossible to squeeze her bulk through the small door the farmer had constructed in the barn, and later he was forced to pierce a door through another wall of the barn to accommodate these burgeoning beasts.

A farmer undertaking a new barn ought to give a lot of thought to how the interior of the building will be accessed. Should the barn have doors large enough allow a haywagon or other farm equipment to enter and exit the barn? Might it also have smaller, people-sized doors? And what about doors for livestock? How do you keep the cows from escaping when the haywagon door is opened? Barn doors can be as simple as vertical supports with a wooden cross-beam to close the door, or as fancy as Dutch doors with ornate wrought iron latches and X-shaped woodwork. They can be rectangular or round-headed, single, double, or Dutch-style.

When it became widely known in the nineteenth century, the rolling door was heralded as one of the most important barn-related inventions in history. Instead of opening outward, the wagon door rolled sideways along a track. This practical design meant that giant barn doors would no longer be blown off their hinges in a tempestuous storm. Nor would they take up a lot of space or get in the way of the mules pulling the haywagon into the barn.

Dutch doors are another practical door design used especially in livestock barns. Dutch doors, secured at the bottom but swinging open at the top, allowed for ventilation while at the same time keeping in the livestock and the wheat on the threshing floor. They also keep out elements.

Along with the carpenter, the blacksmith plays an important role in the barn door. Iron latches, straps, hooks, hinges, and bolts are integral parts of the door. But these forged elements are much more than functional. They are among the most decorative elements of the barn. Door hinges and straps terminate in patterns resembling cloverleaves, arrowheads, snake heads, and spades. Sometimes vegetal forms decorate the metal strapwork that embellishes the barn door. The art of the blacksmith lends a decorative accent to the barn.

On a Louisiana cotton farm, the sloping roof of this red barn seems to hug the cotton field, where the white crops resemble freshly fallen snow.

Oh, farmers, pray that your summers be wet and your winters clear.

—Virgil

THE CORN CRIB

When settlers filtered into the south from Pennsylvania, they turned from growing wheat to cultivating corn crops. Because there was no longer a need for a vast threshing floor, these new Virginia farmers needed only simple log constructions instead of giant barns with vast interior spaces. Called corn cribs, these small log constructions housed harvested ears of corn. Corn needs ventilation in order to dry properly, so the logs were "unchinked," or left with a space between each one so that air could circulate among the ears. These rugged log constructions fit seamlessly into the cozy wooded farms they occupy.

THE LIVESTOCK FEEDER BARN

Sticking to the three-bay English barn prototype, southern farmers developed a specialized kind of barn known as the livestock feeder barn. The central bay was a drive-through opening that could double

This painting evokes daily life on an eighteenth-century farm in Colonial Williamsburg. The barn vies with the farm house for importance, where raising livestock and cultivating crops made for rigorous subsistence farming. ***Colonial Williamsburg Foundation.***

Horse barns are special constructions for special animals. Horses do not tolerate frigid temperatures, so these barns incorporate modern climate-control systems and other high-tech features while still adhering to classic barn architecture.

OPPOSITE
The title of Steve Musgrove's painting, *Shadows Cast, Shadows Past*, reminds us that if only such time-weathered barns could speak, they could tell a thousand stories.

FOLLOWING PAGE
While many barns are painted red or left to weather naturally, in the hot climate of the South some farmers choose light colors that reflect the sun's heat.

I saw the spiders marching through air
Swimming from tree to tree that
mildewed day
In latter August when the hay
Came creaking to the barn

—Robert Traill Spence Lowell

as a work space. One of the flanking sides was used to store farm equipment and other materials, and the remaining side was used to feed and shelter livestock. Once developed, the livestock feeder barn eventually spread north and west to the Corn Belt of the Midwest.

A variation of this type of barn is the dogtrot barn. It has a wide central passage or driveway with a tall roof that accommodates a wagon or a tractor with ease. Flanking either side of the passage are spaces for storage that might also be open to the elements.

The warm southern climate made an enclosed space optional, and livestock feeder barns are sometimes little more than shelters, open to the elements on the sides. Mules, horses, and cows can rest at the tie-ups under the shelter of the long, low aisles on the outer sides of such a barn. Because of the climate, animals do not need to be put in enclosed stalls; they can be tied up under an otherwise open shelter.

In Appalachia, farmers combine the naturally weathered pine and oak of the clapboards with tin or aluminum roofs that amplify the rain

Converted Barns

In 1789, ninety percent of Americans worked on farms and in food production. A century later, still more than half of Americans worked on farms. By 1930, however, only twenty-five percent of Americans were farmers; today, less than five percent of the United States' work frce is made up of farm laborers.

While barns continue to dot the landscape and be seen from the highway, today few of us have ever set foot inside one of these structures. Barns have become obsolete in our society, and they are disappearing rapidly. The dramatically fewer numbers of farm families in America has caused large-scale abandonment and decline of the picturesque barns that characterize the American landscape.

Organizations such as Barn Again! have pledged to combat the disappearance of the barn from the American landscape. There is a new interest in restoring the historic timber-framed barns of yore. Individuals and companies have converted historic barns into unusual homes, restaurants, a variety of stores, antique shops, and community centers. In formerly rural areas of Massachusetts and Connecticut, dairy farms have been successfully and sensitively converted into unique shopping centers with a historic regional flavor.

Many barns would otherwise have been lost if they were not converted into interesting houses that use the structural features of the barn—the loft, exposed beams, wooden floors—as appealing architectural and decorative features of their homes. Former wagon doors become sliding doors; high-tech kitchens replace the spaces where feeding troughs once lay. In many barns-turned-homes, homeowners opt not to plant shrubs or keep a manicured lawn, but rather to maintain the rural, pastoral setting of the barn, thereby preserving the integrity of the structure within its natural setting. Wildflower and vegetable gardens, however, provide a perfect complement to these structures.

Man, barn, and an extraordinary menagerie of animals peacefully cohabitate in the paintings of Mattie Lou O'Kelly.

Into my heart and air that kills
From yon far country blows:
What are those blue remembered hills,
What spires, what farms are those?

—*A. E. Houseman*

in the verdant, rainy hills of the Carolinas and Tennessee. Anyone who has ever been caught in a rain storm and taken refuge in a tin-roofed barn can attest to the magic, almost musical sound that the raindrops make when they pelt these metal rooftops.

THE CANTILEVER BARN

The cantilever barn is unique to the mountainous Appalachian areas of Tennessee and North Carolina. This barn is constructed of rough-hewn logs like a log house, preserving the spirit of the architecture of the early pioneers in its rustic quality. The cantilever barn gets its name from the cantilevered logs that support a large loft that overhangs the lower story. There are over 300 examples of the cantilever barn in eastern Tennessee alone.

In the humid, mountainous areas of the South, where rain is a daily occurrence, the expansive roof of the cantilever barn acts like a giant umbrella sheltering whatever is below. Livestock huddle beneath the great shoulders of the cantilever barn for protection from storms, and bales of hay might also be safeguarded there. The spaces between the logs allow ventilation so that ears of corn can dry.

Weather vanes come in a variety of forms, but animals are among the most popular motifs.

blinding snow falls in horizontal sheets that can send even the most adventuresome farm hand reeling.

THE CORN BELT

The earliest settlers in the Midwest harvested bumper crops of wheat, but chronic invasions of locusts soon wiped out many wheat crops in middle America. While the Far West recovered to excel in wheat production, disenfranchised midwesterners switched to corn, giving rise to the term "Corn Belt" to describe the Midwestern region whose epicenter is southwestern Ohio. Wisconsin farmers originated the saying, "Calm weather in June sets corn in tune." And they should know. Midwesterners are pioneers in corn growing, and the region remains the heart of corn production in America today.

OPPOSITE
The American West promised a new way of life for prospectors and barn builders, who experimented with many architectural forms in the uncharted territory.

FOLLOWING PAGE
Barns often incorporate lean-to additions for a variety of purposes, including sheltering firewood, farm equipment, and small farm animals.

The steeply pitched gable roof of this barn in Washington state is framed against a dramatic rolling landscape. At the end of a day spent grazing in the fields, livestock gather in the enclosed stable space that surrounds the barn.

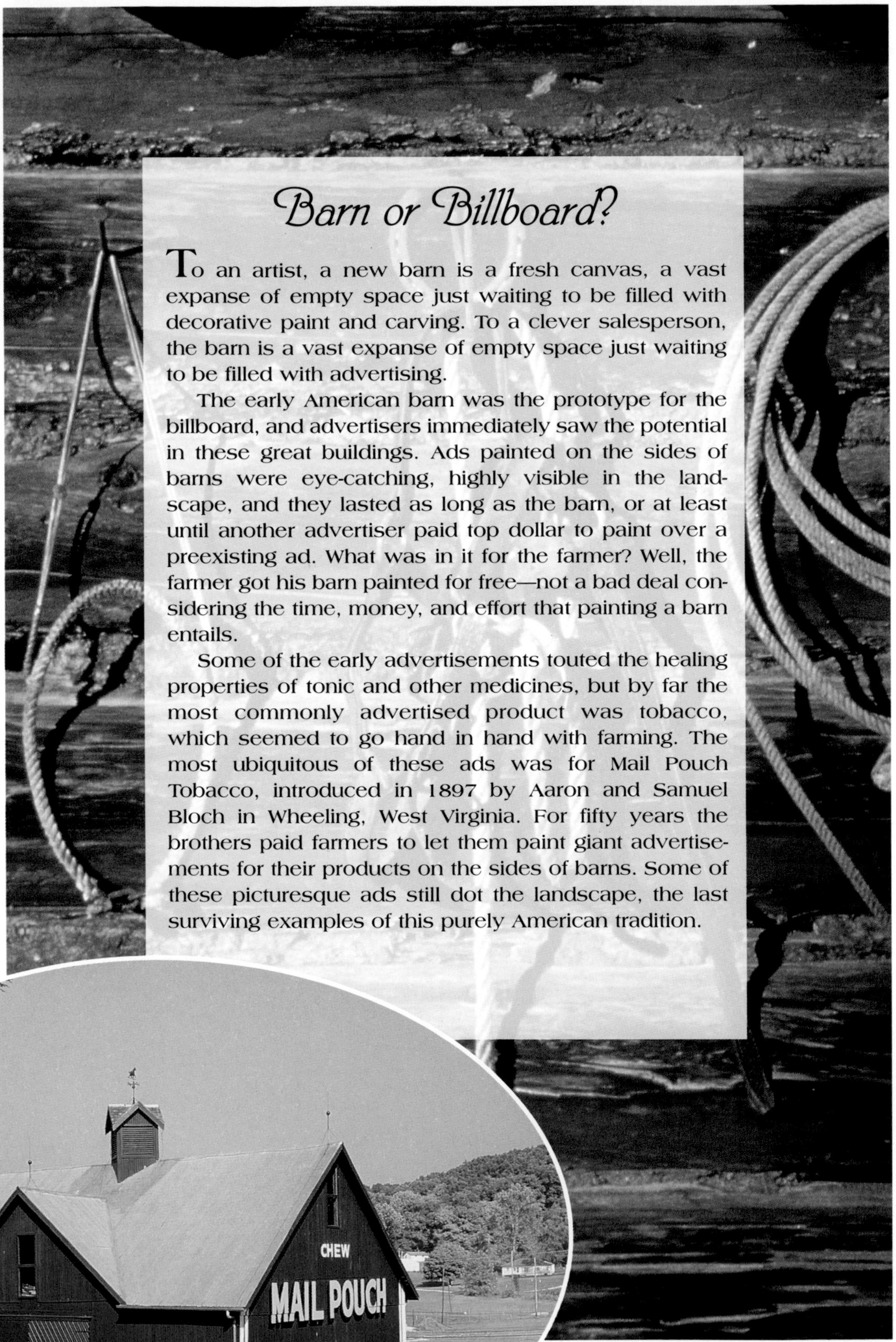

Barn or Billboard?

To an artist, a new barn is a fresh canvas, a vast expanse of empty space just waiting to be filled with decorative paint and carving. To a clever salesperson, the barn is a vast expanse of empty space just waiting to be filled with advertising.

The early American barn was the prototype for the billboard, and advertisers immediately saw the potential in these great buildings. Ads painted on the sides of barns were eye-catching, highly visible in the landscape, and they lasted as long as the barn, or at least until another advertiser paid top dollar to paint over a preexisting ad. What was in it for the farmer? Well, the farmer got his barn painted for free—not a bad deal considering the time, money, and effort that painting a barn entails.

Some of the early advertisements touted the healing properties of tonic and other medicines, but by far the most commonly advertised product was tobacco, which seemed to go hand in hand with farming. The most ubiquitous of these ads was for Mail Pouch Tobacco, introduced in 1897 by Aaron and Samuel Bloch in Wheeling, West Virginia. For fifty years the brothers paid farmers to let them paint giant advertisements for their products on the sides of barns. Some of these picturesque ads still dot the landscape, the last surviving examples of this purely American tradition.

These new corn crops facilitated a major change in farming that would chart a course for Midwestern farmers from then on: They were pioneers in cultivating livestock. Corn is the perfect food for beef, hogs, and cows. Midwestern farmers adapted the livestock feeder barn that had been developed in the South, and the form gravitated northward

Many barns across the country are constructed of wood over a stone or brick foundation. In this historic barn in Iowa, sunset brings sheep and goats back to their stalls on the ground floor.

Arthur F. Tait was renowned for his paintings of the frontiersmen of the wild west, but this intimate portrait of farm animals, entitled simply *Barnyard*, evokes a peaceful country scene.

Barns back east have weather vanes on them to show which way the wind is blowing, but out here there's no need. . . . Farmers just look out the window to see which way the barn is leaning. Some farmers . . . attach a logging chain to a stout pole. They can tell the wind direction by which way the chain is blowing. They don't worry about high wind until the chain starts whipping around and links begin snapping off. Then they know it's likely the wind will come up before morning.

—*Charles Kuralt*

from Tennessee and Kentucky. Farmers stored the grain in giant cribs with open slats similar to the unchinked log corn cribs of the rural South. Midwesterners also constructed impressive dairy farms after the Civil War to meet an increased demand for milk; special barns were set up to facilitate milking year-round.

With the switch to farming based on animals, midwesterners constructed massive barns for corralling cattle herds and gathering hogs. Eventually, they constructed fancy horse barns to accommodate these regal beasts, who do not tolerate extreme cold or heat, and need specialized barns. As midwesterners and westerners turned to ranching, their barns reflect a perfect marriage of form and function in farm architecture.

A VARIETY OF TYPES

As elsewhere in America, the earliest barns in the West were log constructions with thatched roofs. Corner posts were first erected, then the industrious settlers stacked logs on top of one another to create walls. Some of these barns can still be found in the Midwest today. Creating a log barn must have been quite a monotonous task. Legend has it that an early Ohio farmer stacked so much wood in creating his barn that he went mad, and began stacking everything on his farm, obsessively gathering and stacking rocks, bales of hay, and even attempting to stack chickens and piglets around the barnyard.

In contrast to the eastern seaboard, the Midwest and Far West were settled

The vast size of ranches in the Midwest and Western United States require barns with wide side aisles to accommodate livestock. A variety of other outbuildings provide necessary storage.

The Round Barn

The circle has always been a powerful symbol. A circle suggests eternity, something unbroken, complete, and inherently perfect. Well, what about round barns? In the nineteenth century, psychologists and phrenologists promoted the belief that it was mentally and psychologically "healthy" to look at and to be inside round buildings, including barns. Contemplating these round forms was thought to be therapeutic and calming on the mind.

Folkloric tales of the era across America recounted that round barns even had the power to ward off evil. These stories were based on the widely held superstition that the devil could literally "corner you" in a building with angles. There are no corners in a round barn, so people and animals were therefore safe from any demonic activity.

More common than the perfectly round barn is an octagonal variation, with many examples in the Midwestern United States. The octagonal barn is easier to construct than a purely round barn since it uses angles, but the interior space conveys a similar effect. Many of these are supported by a central post or an arcade, creating an outer aisle. Twelve-sided, or duodecagonal barns, are also known but are less common that the octagonal examples. George Washington is said to have designed a sixteen-sided barn in 1793 on his farm in Fairfax County, Virginia, with a large threshing floor in the center and spaces for livestock along the walls.

Utopian societies like the Shakers, embraced the symbolic possibilities of the round barns; the stunning example in Massachusetts' Hancock Shaker Village is the most famous. There are about fifteen round barns remaining in Vermont, but Indiana wins the award for the greatest number of round barns in the United States. Most of its over two hundred examples are located in the central, northern part of the state, and were built largely between 1900 and 1920. Despite its efficient use of space and its aesthetic properties, the round barn failed to win widespread appeal, and remains a rare curiosity in the American landscape. It is sure to turn heads wherever it stands.

by national and ethnic groups that were not as strictly divided as they had been on the east coast in the formative years of the country. Not surprisingly, these farmers used whichever type of barn they liked best, and they invented new hybrid forms of barns, taking the best features from the earlier, more pure designs of the Pennsylvania Dutch and the English.

There are so many variations of barns in the Midwest and West that they defy classification. Wisconsin, known for its cheese, showcases dairy barns that house cows on the ground floor and storage above. Iowa has many livestock feeder barns similar to those in Kentucky and Tennessee. Even typical New England saltbox barns can be found as far west as Washington state. Midwestern farmers embrace the idea of the round barn with great enthusiasm; while they are still considered rare, Indiana boasts over two hundred round barns, more than any other state in the country.

On large dairy farms, the three-bay, two-story barn with storage above and livestock below became standard. As in early Pennsylvania

FOLLOWING PAGE
Midwestern and Western barn builders adapted their designs from a variety of building styles borrowed from the East. The barn on the left recalls the early barns of New England. The one on the right would be equally at home in the hills of Tennessee.

For many at the end of the nineteenth century, the round barn represented a utopian vision.

OPPOSITE
The vast size of ranches in the Midwest and Western United States require barns with wide side aisles to accommodate livestock. A variety of other outbuildings provide necessary storage.

barns, an earthen ramp often gave access directly into the loft for storing crops, hay, or grain. Eventually, hay lofts superseded these banks, and hay could be pulled up into the loft by pulleys that operated hay forks resembling giant metal claws or talons descending to trap the large bales of hay below. At harvest time, the historic barns of the Midwest nearly burst with the bounty of crops that today's farmers produce. But a popular proverb originating in Illinois claims that "the barn is never so full that you can't slap in a bundle at the door."

In the vast prairie of middle America, barn builders conceived of a high-pitched central workspace flanked on either side by very broad side aisles that accommodated stalls for livestock. The roofs slope almost to the ground, giving the impression of an immense bird or an

The old track for the hay fork and a gaping hayloft door distinguish this historic barn in Washington state.

T

airplane with a tremendous wingspan protecting the animals and farm equipment beneath it.

Horse barns are also common in the Midwest and the West. Horses tolerate cold temperatures less well than sheep and cows, and although they have special needs that place horse stables in a separate category, these buildings are closely related to barns. The shedrow is a row of stalls at a racetrack to house horses. Horse barns in the Midwest can be divided up into individual compartments on the inside, with high windows to allow in light. Horse barns are generally much more fancy than other types of barns.

Since the 1960s, corn and soybeans have replaced livestock as the primary interest of midwestern and western farmers. The charming old wooden barns that were built to last forever are not large enough to house the massive high-tech machinery that now dominates farm life in the region. Lightweight but sturdy metal structures are gradually taking the place of these picturesque older structures.

Technological advances in farming make wagon wheels that were once essential to the farm economy picturesque vestiges of our rural past.

Looking to the Future

Technological advances always make life easier. For the American barn, the mid-nineteenth century was a time of profound change. The widespread use of the railroad meant that wood for barns could be shipped all over the country; barn builders were no longer limited to materials native to their regions. Agricultural journals all over America promoted new technical features like cupolas, sliding doors, and silos, which improved the difficult conditions of farm life.

In addition, pre-fabricated barn kits made raising a barn a cinch toward the end of the nineteenth century. These kits could be ordered from a catalog and shipped by rail to nearly any location in the country. While this new way of barn building was undoubtedly convenient, it meant that local diversity in barn architecture was often sacrificed. Farmers in the Pacific Northwest could have New England salt box style barns, and farmers in the South could have perfect examples of the New World Dutch barn.

In the twentieth century, farmers incorporate new technological inventions into their daily lives at a breathtaking rate. Modern barns embrace all the conveniences of modern technology. Most are wired for electricity, running water, and high-technology equipment like milking machines on active dairy farms. Modern barns utilize feed carts, automatic waterers, and mechanisms for carrying out manure on conveyers so that it can be reused as valuable fertilizer for next season's bounty.

Pre-fabricated barns made out of a variety of materials are common. Concrete foundations and steel frames covered by galvanized sheet steel or aluminum can be erected quickly and efficiently, and are also more fire resistant compared to their wooden predecessors. These new barns are several times bigger than the earlier timber structures, allowing room to house the massive machines that make for large-scale farming of the kind our forefathers could have never fathomed. Recently, however, there has been a renewed interest in the venerable old timber structures that grace the pastures of America. In addition to preserving these important structures, many people are also building new timber barns based on the design principles of these relics of our heritage.

A misty sunrise illuminates the skeleton of this old barn, and the eerie dawn light evokes the mysterious qualities of this disused yet timeless landmark in Missouri.

Destroy our farms and the grass will grow in the streets of every city in the country.

—William Jennings Bryan

PIONEER INGENUITY

Letters and diaries of early westward pioneers give us a snapshot of their difficult daily lives, forging a farm in often very inhospitable conditions. From the beginning, midwesterners and westerners have proven themselves among the most pioneering and ingenious of all American farmers, probably due in part to the great challenges they had to face in this new land.

Midwesterners are ahead of their time in the technical revolution that has transformed farming in general and barns specifically over the past 150 years. Iowa farmer William Louden disseminated the monorail hay carrier and the litter carrier, both mechanisms for moving hay into the barn, and moving animal waste out via conveyers. The manure could then be efficiently utilized to fertilize crops. As early as 1824, you could even purchase some pre-fabricated barns in the Midwest. As opposed to subsistence farming, which was the standard in many parts of the country, midwesterners were instrumental in modernizing farming as an industry.

FOLLOWING PAGE
Outlined by a fresh dusting of snow, this historic barn graces a pioneer settlement in Utah. Early westward pioneers faced staggering challenges in this beautiful yet often harsh environment.

AMERICAN ICONS

Grant Wood's famous painting *American Gothic*, which depicts an elderly farm couple staring soberly out of the picture, sums up the grave, unflappable quality of these honest, hardworking, and serious farmers of the Midwest. An Iowa artist, Wood celebrated the rural life of the Midwest. *American Gothic* is an icon of industrious, no-nonsense American farm

This Mona Lisa of the Midwest is just one example of the many Old Master paintings reproduced on the sides of barns throughout America. Inspired by the works of Rembrandt, Michelangelo, and other masters, local artists bring high art to the farm.

Silos

To our modern eyes, the hard steel silo seems almost out of place on a rustic farm, resembling a missile or a rocket ready to blast off into space. Despite its incongruous appearance, however, the silo plays an important practical role on the modern farm. Silos can store tremendous amounts of grain, alfalfa, or other crops. This material, called sileage, is used to sustain livestock over the course of the winter.

Originally, early Americans used barns primarily for grain storage like their European ancestors, and livestock generally were kept outdoors. Eventually, as it became more common for livestock to be housed inside the barn, farmers constructed the silo as a means for storing grain. Corn or other crops are placed inside the silo and tamped down tightly to maximize the interior space. The ability to feed animals all year long from this material allows for more efficient farming.

Silos are a relatively recent development, appearing in the American landscape from the late nineteenth century onward. The earliest silos were not made of metal at all, but were fashioned from vertical wooden boards attached by metal bands to an inner frame. Later silos use a variety of building materials, including concrete, brick, and metal. The silo's characteristic upright cylindrical design became the standard, emerging as a distinctive landmark in the American countryside.

PRECEDING PAGE
A unique wagon-wheel fence leads to this barn in Washington state.

This majestic, sparkling white barn in Michigan is reminiscent of a church. Its distinctive ogee-shaped roof is unusual in barn design, and provides a spacious upper story.

families of the early part of this century. Like the couple in Wood's painting, the barns of the Midwest are also American icons.

Though little, I'll work
as hard as a Turk,
If you'll give me employ,
To plow and sow,
and reap and mow,
And be a farmer's boy.

—*Anonymous*

Afterword

Dwight D. Eisenhower wisely commented that "farming looks mighty easy when your plow is a pencil, and you're a thousand miles from the cornfield." Indeed, barns and farming seem awfully romantic until you've actually tilled the rocky soil or awakened at four A.M. to milk a cow. But great writers and thinkers have always touted farming as one of the most noble professions in the world. In 1895, Booker T. Washington commented that "no race can prosper till it learns there is as much dignity in tilling a field as in writing a poem." A popular saying contends that "a farmer on his knees is higher than a gentleman on his legs."

The barn is the foundation of a farm. The picturesque qualities of American barns lure the traveler to our country's back roads. Extraordinary architectural and historical features both unify these monuments and account for their diversity. Barns have inspired generations of Americans. How many children's nursery rhymes, tales, songs, and fables are set inside the

Barns—even dilapidated ones like this—still serve as gentle reminders of America's vast bounty, and make sentimental photo subjects.

barn? You have only to think of *Charlotte's Web* or the song "Old MacDonald" to realize that the barn has inspired American daily life and culture throughout the centuries.

There is a certain mystique about barns. A man in Georgia salvages wood from destroyed barns in order to makes picture frames for folk paintings like those of Grandma Moses; the bolt holes, vestiges of paint, and weathered patina evoke the spirit of the old barn. In Vermont and New Hampshire, disused barns are often turned into antique shops, drawing people from far and wide in search of treasure. Barns are places to explore, locales where, if you're lucky, you might find a diamond in the rough: a vintage sign, a wonderful rustic pitchfork made of wood, Depression-era glass to add to your collection, a weather vane a little rusty around the edges, or an antique chair just waiting to be restored to its original splendor.

Even though few of us are farmers today, the image of the barn remains a powerful symbol in the American psyche. Fast food restaurants use bright red roofs and incorporate farm animals like chickens and pigs on their neon signs to lure customers off the highway in search of down home country fare. Gambrel roofs, weathered clapboards, and weather vanes lend a rustic air to suburban homes that draw their country flavor from barns. Barns are landmarks, the symbols of rural life that spark our curiosity and fan the flames of the American spirit.

In any landscape, a round barn stands out as an architectural masterpiece.

INDEX

Page numbers in bold-face type indicate photo captions.